Lisiane Fernanda Simeão de Azevedo

Influence of Urea-Formaldehyde Resin on the processing of MDF boards

Lisiane Fernanda Simeão de Azevedo

Influence of Urea-Formaldehyde Resin on the processing of MDF boards

Correlation between different levels of Urea-Formaldehyde Resin and the mechanical strength of MDF boards

ScienciaScripts

Imprint
Any brand names and product names mentioned in this book are subject to trademark, brand or patent protection and are trademarks or registered trademarks of their respective holders. The use of brand names, product names, common names, trade names, product descriptions etc. even without a particular marking in this work is in no way to be construed to mean that such names may be regarded as unrestricted in respect of trademark and brand protection legislation and could thus be used by anyone.

Cover image: www.ingimage.com

This book is a translation from the original published under ISBN 978-613-9-67193-9.

Publisher:
Sciencia Scripts
is a trademark of
Dodo Books Indian Ocean Ltd. and OmniScriptum S.R.L publishing group

120 High Road, East Finchley, London, N2 9ED, United Kingdom
Str. Armeneasca 28/1, office 1, Chisinau MD-2012, Republic of Moldova, Europe
Printed at: see last page
ISBN: 978-620-8-15447-9

SUMMARY

SUMMARY

The aim of this study was to evaluate the use of fibers from the epicarp of babassu coconut in combination, or not, with eucalyptus fibers to make cold-resin wood panels. The panels were pressed at room temperature using 23% of the dry fiber base of synthetic urea-formaldehyde resin. The physical and mechanical properties evaluated were: moisture content (TU), thickness swelling (IE) at 2 h and 24 h, water absorption (AA) at 2 h and 24 h, density and flexural strength (RF), according to ABNT NBR 14810-2. The sheets made from 50% babassu epicarp fibers, combined with 50% eucalyptus fibers retained on the 0.85 mm sieve, obtained the lowest values for swelling in thickness and water absorption. With the 100% babassu sheets and the combined fiber sheets (50% babassu and 50% eucalyptus), the highest values were found in the density test. Using 100% babassu fibers, 100% eucalyptus fibers or a mixture of the two fibers (50% E + 50% B) to make cold-resin panels, it was found that the values obtained performed well. In general, the panels made and tested physically and mechanically have the characteristics of HDF, as they showed results higher than those of the NBR 14810-2 standard. Therefore, babassu coconut epicarp fiber has technical potential for application in cold-cured wood fiber boards.

Keywords: natural fiber, synthetic resin, cold curing.

1   INTRODUCTION

The current trend is to move more and more towards industrialized construction systems, where costs and execution times are absolutely controlled. MDF (Medium Density Fiberboard) is a material made from reforested wood, such as eucalyptus, which, together with metal or wood profiles, as well as other integrated technologies already present in Brazil, makes it possible to create an innovative construction system that is used all over the world, especially in the furniture industry. [3]

MDF is a medium-density wood sheet made from the agglutination of wood fibers with synthetic resins and the combined application of temperature and pressure. To obtain the fibers, the wood is cut into small chips which are then ground by equipment called shredders. [1]

MDF is mainly used in the furniture industry and is often used as a furniture component for parts that require special machining. In the construction industry, it can be used as thin flooring, skirting boards, door pads, partitions and jambs. The reuse of raw materials from waste is a practice that generates a series of social, economic and ecological advantages for the country's development. [8]

The government's incentive in its social programs tends to increase the incentive for the development of projects that use alternative materials that provide lower production costs and that increasingly reduce the problems of burning wood, which greatly increases the amount of solid waste, deserving greater attention to this environmental impact. Among these, the use of Babassu coconut fibers (babassu is a plant from the Palmae family) has shown great interest for use in MDF panels [3].

Brazil's babassu plantations are basically concentrated in the northeast, north and central-west regions, with the northeast region currently having the largest production of almonds and the largest area occupied by babassu coconut.

The current interest in composites with natural fibers is due, among other factors, to the growing worldwide concern with preserving the environment and using renewable raw materials. There is great interest in the search for natural fibers that can adequately replace synthetic fibers. [3]

Therefore, the development of projects that use alternative materials that provide lower production costs, reduce deforestation and reduce the amount of solid waste justifies the execution of this project for the development of the country.

The main objective of this work is to evaluate the influence of the addition of urea formaldehyde resin in the processing of MDF panels based on babassu coconut epicarp fibers added to eucalyptus fibers.

2 OBJECTIVES

2.1 General

To evaluate the influence of the addition of the synthetic resin urea-formaldehyde in the processing of MDF panels based on babassu coconut shell fibers, whether or not added to eucalyptus fibers.

2.2 Specific

- The samples were made using 23% Ureaformaldehyde resin in relation to the total dry fibers;

- Produce MDF samples using a combination of 50% eucalyptus fibers and 50% babassu coconut;

- Produce MDF samples using 100% babassu fibers;

- Produce MDF samples using 100% eucalyptus fibers

- Analyze the physical and mechanical properties of the samples produced;

- Compare the data obtained through the research with the parameters adopted by Brazilian technical standards (ABNT-NBR 15316) for "Medium or high density fiberboard."

3 THEORETICAL BASIS

3.1 Wooden Panels

Wood is one of the oldest natural resources used by man in his daily activities. However, the industrialization process between the end of the 19th century and the beginning of the 20th century resulted in a decrease in good quality wood and an accumulation of waste generated from its processing, such as sawdust, chipper chips, blade and roller refills, which were thrown away or at most used as fuel. From the end of the 19th century, white eucalyptus was introduced into Spain, Portugal, Chile, Ecuador, Uruguay, the United States, India, Morocco and Ethiopia, among others. [2]

In the United States in the 1950s, there was little use of felled trees by the end consumer and the advent of new technologies was developed with the aim of minimizing waste and gradually this material was reused, reaching a use rate of over 90% of felled logs. [5]

In line with global trends, the Brazilian panel industry has experienced significant growth in recent years. According to the BNDES sector study in 2010 on "Wood Panels in Brazil - Market Overview: Wood Panels", ABIPA expects demand for reconstituted wood panels to grow by an average of 14.1% p.a. between 2009 and 2013, exceeding growth projection expectations, and as a consequence, an increase in demand for reforested pine and eucalyptus wood. [4]

3.1.1 Wood fiber panels

Wood fiberboard is a product made from the refinement of lignocellulosic fibers, with primary adhesion occurring through the interlacing of the fibers and the adhesive properties of some of the wood's chemical components, such as lignin.[14].

3.1.1.1 Classification of wood fiber panels

Fiberboard is usually classified by its density and can be dry-processed or wet-processed. Dry processing is applicable to boards with high density (hardboard) and medium density (MDF). Processing can be carried out with or without a pressing stage. Table 3.1 below shows the types of wood fiber panels.

Table 1 - Classification of Wood Fiber Panels

Type	Density g/cm3
Not pressed	
1. Insulated panels (semi-rigid)	0,02 a 0,15
2. Insulated panels (rigid)	0,15 a 0,40
Pressed	
1. Medium Density Fibreboard (MDF)	0,5 a 0,8
2. Hard fiberboard	0,8 a 1,20
3. High Densification Panels	1,20 a 1,45

Source: [8]

3.2 Raw materials

3.2.1 Eucalyptus fibers

Natural fibers are classified as vegetable, animal or mineral. The most prominent natural fibers are vegetable fibers due to the wide variety of plants found in nature today.

Plant fibers came to prominence in Brazil in the 1990s, when they began to replace synthetic fibers, confirming their lower degrading power compared to synthetic fibers. Currently, plant fibers are of great importance as reinforcements in the manufacture of polymer composites.

The main advantages of this type of fiber are its abundance in nature, easy access to fibers, low acquisition costs associated with low density and the fact that they are renewable and biodegradable materials. [17]

The history of eucalyptus in Brazil shows that the plant was introduced to the country in the 20th century as a raw material for railroad construction, and was only later used in the pulp and furniture industries. The origin of this plant, which belongs to the myrtle family, was in Australia and there are currently more than 700 catalogued species, including: *Eucalyptussaligna, E. grandis and E. urophila (and its hybrid, E. urograndis)*, which are best known in Brazil. [21]

3.2.2 Babassu Coconut Fibers

As a native Brazilian plant, the babassu palm is important in the natural cycle of soil conservation and improvement, as a successional forest from capoeira (or deforestation) to large forest. This plant is being valued on the national and regional market, mainly due to the total utilization of some of its parts. Among its general characteristics, it has a monocule that measures up to 30 meters in height and a brown fruit that can measure between 8 and 15 cm or even between 5 and 7 cm in diameter. These characteristics help define the harvest period. Each year, a babassu palm can produce up to 2000 fruits, which weigh between 90 and 280 grams. In Brazil, it is found in the North, Northeast and West regions, with the largest numbers in the states of Maranhao, Piaui, Cearà, Pernambuco and Alagoas. Outside the country, this plant can also be found in the South American countries of Bolivia, the Guianas and Suriname. [20]

For a long time, the importance of babassu in the Brazilian market was focused solely on the kernel and the use of the endocarp, mainly due to the lack of knowledge of other possibilities for exploiting parts of the fruit such as the mesocarp and epicarp. In recent years, however, there has been a drop in the plant's productivity, caused, among other things, by a lack of investment in coconut extraction and in valuing the professionals who provide labor during the collection and breaking of the coconut.

Babassu is economically important for Brazil, it is among the products that stood out in the country in 2009; the first were açai fruits (R$ 160.5 million), then babassu almonds (R$ 121.3 million), piassava fibers (R$ 110.3 million), native yerba mate (R$ 86.6 million), carnauba wax powder (R$ 79.4 million) and Brazil nuts (R$ 52.2 million). All these products together accounted for 89.1% of non-timber forest production[4].

Up to 64 products can be extracted from the plant, such as methanol, fatty acids, glycerin and flour. The fruit can be used 100% of the time, and the most important industries are those that produce food, perfumes and/or cosmetics. [6]

3.2.3 . Resins

The main adhesives used in the production of these panels are urea formaldehyde and melamine formaldehyde. According to this researcher, urea-formaldehyde resin is an adhesive used in more than 90% of wood panels, due to its low cost compared to other available resins. It is a light-colored, non-flammable adhesive, can be used internally where there is no humidity, and can be cured either hot or cold. [18]

Urea formaldehyde resin is the most widely used binder in the manufacture of panels such as chipboard and MDF. The widespread use of this resin in the manufacture of panels is due to its low cost.

3.2.4 Medium-density fiberboard (MDF)

The Composite Board Association of the United States defines MDF (Medium Density Fiberboard) as a dry-formed board made from lignocellulosic fibers combined with a synthetic resin or other binding agent, compacted by hot pressing (density of 496 to 801 Kg/m^3) in a process in which the totality of the adhesion between the fibers depends on the adhesive added. [14]

MDF sheets can be classified for marketing purposes into (ABNT STANDARD NBR 15316-1, 2006): [16]

a)   HDF: sheet with a density > 800 kg/m^3 ;

b)   Standard: sheet with a density of > 650 kg/m^3 and < 800 kg/m^3 ;

c)   Light: sheet with a density of < 650 kg/m^3 ;

d)   Ultra light: sheet with a density < 550 kg/m^3 .

3.2.5 Physico-mechanical properties required by MDF

MDF manufacturers in Brazil use the main national standards to carry out product characterization in in-house laboratories. However, the main product-related requirements that must be met are equivalent to those presented in the European EN Standards. The following definitions for the tests are mentioned below [11]:

a)   Water absorption: Increase in the mass (in water) of a medium-density fiberboard (MDF) specimen after being immersed in water at $(20 \pm 1)°C$ for 24 h $\pm$ 36 min;

b)   Swelling: Percentage variation in the increase in thickness of a medium-density fiberboard (MDF) specimen after being immersed for 24 h $\pm$ 36 min. in water at a temperature of $(20 \pm 1)°C$;

c)   Density: Characteristic represented by the ratio between the mass and volume of a body at a given moisture content;

d)   Moisture content: The percentage of water released from a medium-density fiberboard

(MDF) specimen when it is subjected to a temperature of (103 ± 2) °C until the mass becomes constant. The percentage is taken on a dry basis;

e) Density profile: Gradient that determines the density profile of an MDF sheet specimen by partially determining the density of its layers, classifying the MDF sheet according to its final density level;

f) Static bending strength: The strength of a medium-density fiberboard (MDF) specimen, supported at its ends, when subjected to a force applied at its center until it breaks;

g) Resistance to perpendicular traction (internal bond): Resistance offered by a medium-density fiberboard (MDF) test specimen when it is subjected to a traction force applied perpendicular to its surface until it breaks.

The minimum property requirements according to the values proposed in the Brazilian Standard and the European Standard for MDF panels are described in Tables 3.2 and 3.3, respectively.

Table 2 - Main requirements according to ABNT NBR 15316-2

Properties	Unit	Thicknesses 12 to 19 mm
Swelling 24 h (maximum)	%	12
Perp. tensile strength (min.)	N/mm^2	0,55
Flexural strength (Max.)	N/mm^2	20
Modulus of Elasticity	N/mm^2	2200

Source: Adapted from Table 2 of ABNT NBR 15316-2[16]

Table 3 - Main requirements according to EN622-5

Properties	Test Method	Unit	Thicknesses 12 to 19 mm
Swelling24h (maximum)	EN 317	%	12
Tensile strength Perp (min.)	EN 319	N/mm^2	0,55
Flexural Strength (Max.)	EN 310	N/mm^2	20
Modulode Elasticity	EN 310	N/mm^2	2200

Source: Adapted from Table 3 of EN 622 - 5 [11]

4   MATERIALS AND METHODS

4.1 Materials

The 7-year-old Eucalyptus grandis tree trunks used in this research were obtained from the farm located in the municipality of Urbano Santos, state of Maranhao, belonging to Suzano Papel e Celulose. Rings measuring approximately 12 centimeters were cut at the Suzano company itself and kept in a laboratory oven with forced ventilation at 50 ± 2 °C for 48 hours and an equal period for cooling and humidity stability of around 20%. The rings were also reduced into chips at the company itself by cutting them with the cutting element (cleaver) of a particle chopping machine in an axial direction (in the direction of the fibers). Figure 1 shows the physical appearance of the eucalyptus chips obtained.

Figure 1 - Eucalyptus chips

As a raw material, babassu coconut epicarp was obtained from the company Florestas Brasileiras S.A., located in the municipality of Itapecuru Mirim, in the state of Maranhao. The fibers of the babassu coconut shell (epicarp) were obtained by first grinding the shells in the rotary grinding machine of the company supplying this raw material. They were then dried in an oven at 60° C for 24 hours with forced ventilation and a moisture content of 60%. Figure

2 shows the physical appearance of the babassu coconut shells.

Figure 2 - Babassu Coconut Epicarque Chips

4.2 Methods

4.2.1 Basic Density of Eucalyptus and Babassu Coconut Chips

The basic density of the chips was determined using the immersion method, which was carried out in accordance with standard NBR 14660. To do this, 200g of the materials were collected, chosen at random, with three repetitions for each type of chip, making a total of six samples. These were then placed in a nylon net and immersed in water until they were completely saturated. The mass of the basket and chips as a whole was determined, as was the mass of the basket immersed in water. After this, the disks were placed in an oven at 105 °C to dry until they reached a constant weight. The density was then calculated according to the following equation 1:

$$D_b = \frac{M_3}{M_2 - M_1}$$

Equation 1 - Basic Density

Where:

D_b = basic density.

M_3 = oven-dried mass at $(105 \pm 3)^0$ C, in grams.

M_2 = weighing mass of the immersed assembly (basket and chips), in grams.

M_1 = mass of the immersed basket, in grams.

4.2.2 Generation of Eucalyptus and Babassu Epicarp Fibers

To facilitate the grinding process, the chips were treated using the following procedures: [10].

÷   Immersion in water at room temperature for 72 hours;

÷   Change the water every 24 hours;

÷   Air dry for 48 hours;

÷   Immersion of the fibers in water at a temperature of 80^0 C for 90 minutes;

÷   Immersion of the fibers in sodium hydroxide solution for 72 hours.

After the final water immersion stage, the eucalyptus chips and babassu epicarp were immersed in a 15% sodium hydroxide solution. Figure 3 shows the physical appearance of the chips subjected to this treatment stage with only water and sodium hydroxide solution. (a) Eucalyptus chips treated with water and (c) treated with 15% sodium hydroxide, (b) babassu epicarp fibers treated with water and (d) treated with 15% sodium hydroxide.

Figure 3 - Chips treated with water and 15% NaOH

The experiment consisted of 02 formulations and three repetitions, totaling 6 panels, where the types of panels and fiber mixtures and the different fiber sizes were analyzed, as shown in Table 4.

Table 4 - Panel Formulation

Formulation	Type of fiber	Mixing ratio (%)	Sieve Classification (ASTM)	Size range (mm)	Resin content (%)
P1	Eucalyptus + Babassu	50/50	-16 +20	0,85 - 1,18	23
P2	Eucalyptus + Babassu	50/50	-20(through) +30(retained)	0,6 - 0,85	23

Figure 4 shows the image of the chopped fibers obtained in the chopping mill with two fixed hammers and two knives, as they were received at IFMA.

Figure 4 - Eucalyptus fibers (a) and Babassu coconut fibers (b)

4.2.3 . Resin

The binding material used was a urea-formaldehyde-based resin called Colamite, manufactured by EUROAMERICAM and marketed by LÉO MADEIRAS, manufacturers of MDF panels. The dosage of resin used in this work was 23% of the total dry fibers. The resin content was calculated as follows, as shown in equation 2.

$$TR = \frac{MR}{MF.(1-TU)} . TSR$$

Equation 2 - Resin content

Where:

TR = resin content;

MR = resin mass;

MF = fiber mass;

TU = moisture content;

TSR = resin solids content.

4.2.4 Moisture Content Determination

After the particle size test and before pressing, the particles were dried in a conventional oven at 105°C until they reached a constant mass. The moisture content of the particles was determined by the loss of mass obtained after the drying process in the oven.

4.2.5 Drying the fibers

The eucalyptus and babassu coconut fibers were completely dried in a Quimis mod. QM-85 forced-air chamber at a temperature of 70°C, for a period of time to obtain a constant mass and a final humidity of approximately 3%.

4.2.6 Preparing the fiber mattress

Initially, the glue and fibers were placed in a PAVITEST planetary mixer, used to homogenize the fibers with the resin.

After homogenization, the fibres moistened with the glue were manually compacted in a wooden mould in the shape of a box measuring 30 x 20 x 15 cm to form the particle mattress. Figure 4.5 shows the fiber mattress being pre-formed.

Figure 5 - Wooden mold with Eucalyptus and Babassu fibers

4.2.7 Pressing

The panels were pressed in a SCHWING SIWA equipamentos industriais Ltda manual press, with a maximum load capacity of 15 tons, at room temperature (cold), a compacting pressure of 2 MPa and a pressing time of 8 hours per panel. After pressing, the panels were kept at room temperature for 24 hours to reach equilibrium humidity. They were then squared off and placed in an air-conditioned room with a temperature of $21 \pm 3°C$ and a relative humidity of $65 \pm 5\%$ to stabilize at the appropriate moisture content. Figure 6 shows the panel after compaction in the cold press.

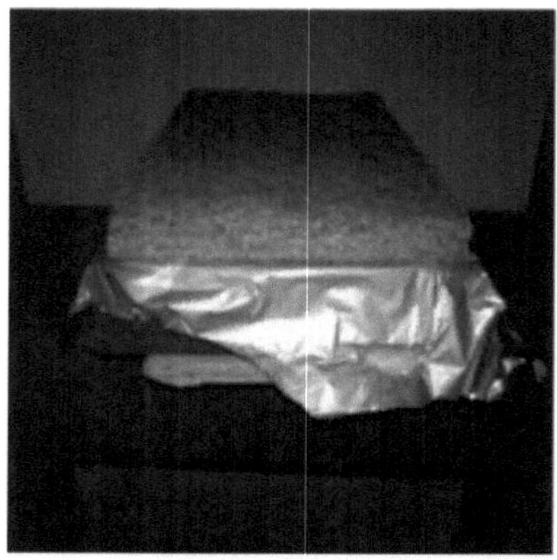

Figure 6 - Fiber mattress

4.2.8 Evaluation of Physical and Mechanical Properties

After the panels had acclimatized, specimens were taken to the IFMA carpentry workshop for the physico-mechanical tests. Three panels of each formulation were made to evaluate the physical-mechanical properties. For each panel, a sample was taken for the flexural modulus of rupture (MOR), totaling 3 samples. For mass density, 3 samples were taken from each panel, totaling 9 samples. For moisture content, 3 samples were taken per panel, totaling 9 samples. For swelling and water absorption at 2 and 24 hours, 3 samples were taken per panel, totaling 9 samples.

4.2.8.1   Tests to determine mass density

Mass density is measured by the ratio of mass to volume in specimens after stabilization in an air-conditioned room at 21°C and 65% humidity. After weighing on a Marte Balanças model MB 500 digital scale, the width and length measurements were determined using a digital caliper with a resolution of 0.01 mm. The thickness was determined using a digital micrometer at 3 points on the specimen with a precision of 0.001 mm, one in the center and the others at the corners of the samples.

These measurements were carried out using the methodology described in standard NBR 14810-2 on samples measuring 50 x 50 x 14 mm. Density was calculated according to the equation below.

$$D = \frac{M}{V} . 1000000$$

Equation 3 - Density

To calculate the volume, the equation was used:

$$V = C . L . E$$

Equation 4 - Volume

Where:

D: mass density (g/cm^3);

M: mass (g);

V: volume (cm^3).

4.2.8.2   Test to Determine Moisture Content

To determine the moisture content, specimens measuring 50 x 50 mm were weighed individually and then placed in an oven with forced air circulation at a controlled temperature of 103^0 C. The moisture content was calculated using the following equation 5.

$$Tu = \frac{M_U - M_S}{M_S} . 100\%$$

Equation 5 - Moisture Content

Where:

Tu = moisture content (%).

Mu = wet mass (g).

Ms = dry mass (g).

4.2.8.3   Thickness Swelling Test

The thickness swelling determinations were carried out after immersion in distilled water with a pH of 7 ± 1, submerged 25 ± 5 mm from the surface at a temperature of 20 ± 1°C for 2h and 24h ± 36 min. To determine this, the specimens were weighed on an analytical balance and their thickness was measured in digital micrometers, with a resolution of 0.001 mm, before and after immersion in water. Thickness swelling was calculated according to the NBR 148102 standard on 50 x 50 mm samples using equation 6.

Equation 6 - Thickness Swelling

$$IE = \frac{E_F - E_I}{E_I} \cdot 100$$

Where:

IE: swelling in thickness (%)

Ef: final thickness (mm)

Ei: initial thickness (mm).

4.2.8.4   Test for Determining Water Absorption

Test specimens measuring 50 x 50 mm were used to determine water absorption (AA), in accordance with NBR 15316. The specimens were immersed in distilled water with a pH of 7 ± 1 and their dimensions and mass were determined after 2 h and 24 h of immersion.

Water absorption was calculated according to equation 7 below:

$$Aa = \frac{M_U - M_S}{M_S} . 100$$

Equation 7 - Water Absorption

Where:

Aa: water absorption (%)

Mu: wet mass (g).

Ms: dry mass (g).

4.2.8.5   Tests to determine static bending strength (MOR)

Using specimens measuring 20 times their nominal thickness in length and width of 50 ± 1 mm, the modulus of rupture (MOR) was calculated from the static bending test. The dimensions of the specimens were measured using a digital caliper. To carry out this test, the 250 x 50 mm specimens were placed on two supports of the EMIC DL 30000 universal testing machine.

Equation 8 below was used to calculate the modulus of rupture (MOR).

$$MOR = \frac{3.F.L}{2.b.t^2}$$

Equation 8 - Modulus of Rupture

Where:

MOR - Modulus of resistance in bending (MPa);

L: distance between supports (mm);

b: width of the specimen (mm);

t: thickness of the specimen (mm);

F: load at break (N).

5 Results and discussion

5.1 Characteristics of fibers

5.1.1 Fiber Granulometry

The grain size of the fibers generated from eucalyptus wood and babassu epicarp residue, under laboratory conditions, is shown in Table 5 and Figure 7.

Table 5 - Granulometry of Fibers Used in Panel Production

Sieve Set (mm) (ASTM)		Eucalyptus fibers (%)	Babassu Coconut Fibers (%)
2,00	10	12,48	12,75
1,18	16	38,30	14,11
0,85	20	25,47	14,18
0,60	30	13,28	23,44
Fund	Fund	10,19	34,61

Source: Analysis of fiber granulometry by student Lisiane de Azevedo.

The results presented in Table 5 show that the eucalyptus fibers provided particles with a larger particle size than the fibers obtained from babassu coconut, given the greater amount of fines generated at the bottom by the babassu coconut fibers.

When comparing the particle size distribution of eucalyptus fibers with those of babassu coconut, there are big differences in the retained masses, especially those retained by the ASTM 16, ASTM 20 and ASTM 30 sieves, as shown in Table 5. However, a greater number of fibers were retained on the ASTM 16 and ASTM 20 sieves for eucalyptus fibers, while a smaller number of fibers were retained on these same sieves for babassu coconut fibers.

However, in the ABNT 30 sieve, the largest amount of fiber retained was babassu coconut fiber. At the bottom, the largest amount collected was babassu coconut fiber, i.e. the largest amount of "fines". The higher percentage of fines (34.61%) generated in the processing of these fibers can cause inadequate adhesion between the particles, since the fines increase resin consumption, which can harm the quality of the panels. Therefore, this portion was discarded in this study.

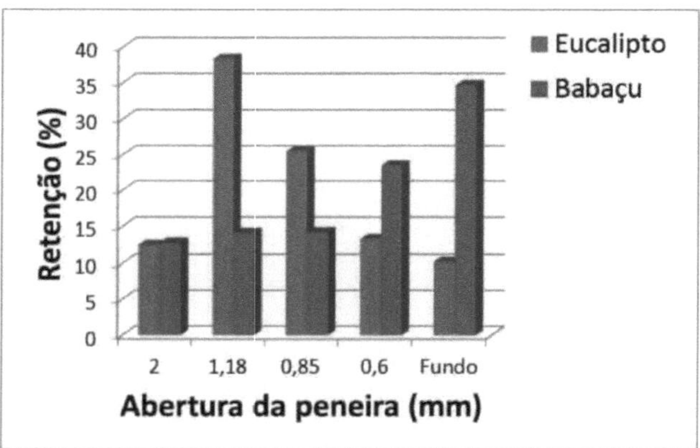

Figure 7 - Granulometry of fibers used in panel production

5.1.2 Moisture content of fibers

The moisture content of the different fibers, before and after applying the resin, is shown in Table 6.

Table 6- Average Particle Moisture Content

Particles	Average moisture content (%)	
	Before resin	**After the resin**
Eucalyptus	$16,41 \pm 0,7$	$2,70 \pm 0,9$
Babassu epicarp	$15,74 \pm 0,4$	$2,61 \pm 0,7$

Source: Moisture content analysis by student Lisiane de Azevedo.

Table 6 shows that the moisture content of the particles before applying the resin is above the range normally used for panel production. This range varies between 3% and 6% moisture. [15]

However, these values were reduced after the drying process to values close to those recommended in the literature, as shown in Table 6.

In panels that are processed using heat, it is necessary to control the moisture content of the particles so that excess steam during the pressing process does not cause defects in the panel. On the other hand, too low a moisture content can cause poor bonding and therefore reduce the mechanical properties. As the processing used in this work was cold pressing, the moisture content used did not compromise the quality of the panels processed with either eucalyptus or babassu fibers. Therefore, defects that can be attributed to hot processing without strict control of fiber moisture content do not apply to the methodology adopted in this work. This is of great importance for reducing panel manufacturing costs by not using the heating stage in processing.

5.1.3 Basic fiber density

Table 7 shows the basic density of eucalyptus and babassu coconut chips before the resin application stage.

The average values obtained were 0.35 g/cm^3 and 0.41 g/cm^3 for eucalyptus and babassu coconut fibers, respectively, in the pre-shredding conditions. It can be seen that the density of babassu chips was slightly higher than that of eucalyptus. Consequently, for the same volume of pressing matrix, the same resin content and the same fiber mass, the eucalyptus chip will occupy a larger volume than the babassu coconut chip. Since the same pressing conditions (mold size and compacting pressure) were maintained in this study, the panels made with eucalyptus fibers were thicker than those made with babassu coconut fibers.

Table 7 - Basic Density of Eucalyptus and Babassu Coconut Fibers

Fiber	Basic Density (g/cm3)
Eucalyptus	0,35 ± 0,01
Babassu coconut	0,41 ± 0,03

Source: Basic fiber density analysis by student Lisiane de Azevedo.

The basic density values found for eucalyptus chips were lower than the 0.43 g/cm^3 standard. This may possibly be related to the different plantation sites and climatic differences in our region compared to the region where the researchers collected their samples. [15]

5.2 Panel properties

5.2.1 Physical properties

5.2.1.1 Moisture content

The average moisture content and coefficient of variation (CV) of the panels after the acclimatization period are shown in Table 8.

Table 8 - Moisture Content of Panels

Formulations	Moisture content	
	Average value (%)	Standard Deviation
P1	10,10	0,24
P2	10,38	0,22
P3	10,48	0,18
P4	10,26	0,34
P5	10,17	0,44
P6	9,96	0,37

Source: Analysis of panel moisture content by student Lisiane de Azevedo.

P1 = eucalyptus retained at 1.18 mm; P2 = eucalyptus retained at 0.85 mm; P3 = eucalyptus retained at 0.6 mm; P4 = babassu retained at 0.6 mm; P5 = mixture of babassu and eucalyptus retained at 0.85 mm; P6 = mixture of babassu and eucalyptus retained at 0.6 mm.

As can be seen from the results in Table 8, the moisture content of all the formulations evaluated complies with the requirements of Brazilian Standard ABNT NBR 15316-2 and consequently its corresponding European Standard EN 622-5, which set a minimum moisture content of 4% and a maximum of 11%.

Studying the use of tauari wood waste (Couratarioblongifolia) in the production of particleboard using castor oil-based resin (*RicinuscommunisL.*), Santos reported in his work that the moisture content of the particleboard was in the range of 5% to 11% as specified by ABNT standard NBR 14810. [19]

Weber, using plywood, MDF and MDP waste to make agglomerated panels, obtained values ranging from 7.34% to 8.96% for the moisture content. Hillig, Haselein and Santini (2004) found values between 3.74% and 8.78% for Pinus elliottii, Eucalyptusgrandis and Acaciamearnsii flakeboards [22].

Table 8 shows that the P6 formulation, i.e. the formulation containing 50%-p of each of the fibers (eucalyptus and babassu coconut) passed through the 0.6 mm sieve, had the lowest moisture content (9.96). However, formulation P3, formulated only with eucalyptus fibers retained in 0.6 mm, had the highest moisture content (10.48).

However, the statistical analysis applied to the average moisture content values for all the formulations shows that there was practically no effect of the type of fiber on this property, regardless of the size of the fiber used.

5.2.1.2 Mass density of the panels

Mass density is a property that is directly linked to physical and mechanical performance, both in solid wood and in processed wood-based panels.

The average values for the mass density of the panels are shown in Table 5.5. The average density values shown in Table 9 ranged from 0.86 g/cm^3 (P1 and P3) to 0.99 g/cm^3 (P4). Analysis of the average density values shows a difference between the densities of panels made with eucalyptus fibers and those made with babassu coconut fibers. It was observed that the panels containing only eucalyptus fibers (P1, P2 and P3) had lower densities than the panels containing only babassu coconut fibers (P4) or when mixed with eucalyptus (50%-p) (P5 and P6). This was due to the lower basic density of the eucalyptus fiber, which occupied a greater volume of material within the matrix and consequently increased the thickness of the sheet, as described above.

Table 9 - Density of Eucalyptus and Babassu Coconut Fiber Panels

Formulations	Panel density (g/cm)^3	
	Average Value	Standard deviation
Pi	0,86	0,13
P2	0,85	0,i6
P3	0,86	0,15
P4	0,99	0,08
P5	0,94	0,10
P6	0,90	0,09

Source: Density analysis of the panels by student Lisiane de Azevedo.

P1 = eucalyptus retained at 1.18 mm; P2 = eucalyptus retained at 0.85 mm; P3 = eucalyptus retained at 0.6 mm; P4 = babassu retained at 0.6 mm; P5 = mixture of babassu and eucalyptus retained at 0.85 mm; P6 = mixture of babassu and eucalyptus retained at 0.6 mm.

It can also be seen that the formulations using only babassu coconut fiber or its mixture with eucalyptus fiber had the highest absolute averages for density in this experiment. This may probably have been due to the greater adherence of the resin to the babassu coconut fiber, thus promoting a greater bond between these fibers, consequently promoting greater densification.

It is important to note that the density results of the panels are of great importance due to their direct influence on the physical and mechanical properties of the panels. The average density is one of the factors that most influences the final properties of the panels. [10]

It is worth noting that in this work the fibers were compacted to form the panels in a manual cold press.

5.2.1.3 Water absorption and swelling in thickness

The average values of the water absorption and thickness swelling properties after 2 hours and 24 hours immersion in water are shown in Table 10.

Table 10 - Water Absorption and Swelling in Panel Thickness

PANELS	AA - 2 h (%)	AA - 24 h	IE - 2 h	IE - 24 h
P1	17,56 ± 2,18	29,14± 2,02	5,74±0,92	10,62±0,63
P2	10,16 ±3,70	20,68±3,05	4,63±2,07	8,76±2,08
P3	7,93±2,58	16,84±2,15	4,10±1,28	7,18±2,25
P4	10,22±3,04	22,53±3,58	3,05±1,32	8,85±0,54
P5	5,71±3,05	15,60±3,35	2,47±1,42	4,06±2,60
P6	15,16±2,79	33,48±3,49	4,39±0,68	10,66±0,51

Source: Analysis of the panels by Lisiane de Azevedo

P1 = eucalyptus retained at 1.18 mm; P2 = eucalyptus retained at 0.85 mm; P3 = eucalyptus retained at 0.6 mm; P4 = babassu retained at 0.6 mm; P5 = mixture of babassu and eucalyptus retained at 0.85 mm; P6 = mixture of babassu and eucalyptus retained at 0.6 mm.

The average values shown in Table 10 for water absorption after 2 hours of immersion ranged from 5.71% (P5) to 17.56% (P1). Water absorption after 24 hours of immersion ranged from 15.60% (P5) to 33.48% (P6). It can be seen that the P5 formulation was the one with the lowest water absorption values both at 2 hours and 24 hours of immersion, consequently the best results compared to the other panels. Panels P1 and P6, on the other hand, showed the least satisfactory results in terms of the water absorption of their panels, both at 2 hours and 24 hours.

The water absorption results show that babassu coconut fiber had an influence on reducing water absorption, with lower water absorption values for panels made with a mixture of fibers, compared to panels containing only eucalyptus fiber (when compared to panels with the same grain size).

The particle size also seemed to influence the water absorption of the panels containing only eucalyptus fibers and also in the formulations containing babassu coconut fibers. However, the particle size, specifically the smaller size of the fibers, had a more pronounced influence on the water absorption of the panels containing the mixture of eucalyptus fibers and babassu coconut fiber (P5 compared to P6).

Comparing the results in Table 10 with the values obtained in other studies, we can see that the water absorption values found in this study were lower than the values obtained by Iwakiri et al (2005). [13]

The thickness swelling results showed similar trends to the water absorption results. The results shown in Table 10 for thickness swelling show that, after 2 hours of immersion, the average values ranged from 2.47% (P5) to 5.74% (P1). Thickness swelling after 24 hours of immersion ranged from 4.06% (P5) to 10.66% (P6). This variation in thickness swelling was practically identical to that found for water absorption.

Evaluating the values presented in the 24-hour water absorption tests, all the panels were within the limits established by EN 317 (1993). As for the thickness swelling results, all the formulations were within the maximum criteria of NBR 15316 (2006) of 12%.

In the work published by Iwakiri et al. (2005), using panels produced pure or in mixtures from sawmill residues of three species of Eucalyptus (E. maculata, E. grandis and E. tereticornis), with a nominal density of 0.75 g/cm and two proportions of UF resin (8 and 50%).

tereticornis), with a nominal density of 0.75 g/cm³ and two proportions of UF resin (8 and 12%), values between 37.37% and 50.80% were observed for absorption and 23.51% and 38.81% for swelling in thickness, both after 24 hours immersion in water. [12][13]

Iwakiri et al. (2005), in their research with Pinus spp. particleboards with a density of 0.650 g/cm3 and 8% UF resin, reported that the values obtained for water absorption and swelling in thickness were 126.58% and 29.99%, respectively, after 24 hours of immersion in water. The same authors found values of 105.79% and 65.84% for water absorption and 21.01% and 26.86% for thickness swelling, respectively, for panels with 0.650 g/cm3 and 0.900 g/cm3, both with 8% melamine urea formaldehyde (MUF) resin[12][13].

Comparing the values in Table 10 with those found by the researchers mentioned above and others, it can be seen that the panels made from babassu coconut epicarp fibers performed well in terms of water absorption and thickness swelling.

5.2.2 Mechanical properties

5.2.2.1 Modulus of Rupture in Static Flexion (MOR)

Table 11 shows the average results for the Modulus of Rupture (MOR) obtained from the mechanical tests in static bending.

Table 11 - MOR Results in Static Bending for the Panels

Formulations	MOR (Mpa)	Standard Deviation
P1	23,09	2,4
P2	3i,i6	3,6
P3	29,83	3,2
P4	29,45	i,4
P5	29,08	2,7
P6	28,60	3,5

Source: Analysis of Lisiane de Azevedo's panels

P1 = eucalyptus retained at 1.18 mm; P2 = eucalyptus retained at 0.85 mm; P3 = eucalyptus retained at 0.6 mm; P4 = babassu retained at 0.6 mm; P5 = mixture of babassu and eucalyptus retained at 0.85 mm; P6 = mixture of babassu and eucalyptus retained at 0.6 mm.

The average values for Modulus of Rupture shown in Table 11, regardless of the type of formulation, ranged from 23.09 MPa (Pi) to 31.16 MPa (P2).

Comparing the panels with eucalyptus fibers, it can be seen that the particle size influenced the mechanical strength of the panels. The MOR increased when the size of the eucalyptus fiber in the 2.0 to 1.18 mm range decreased to the 1.18 to 0.85 mm range. However, there was no significant variation when the fiber size decreased to the 0.85 to 0.6 mm range.

With regard to babassu coconut fiber, the MOR results were practically not influenced by the size of the fibers. It can be seen that the values found for these formulations were just below and very close to the highest value found for the formulation containing only eucalyptus fiber (P2). However, comparing the MOR result of the formulation containing only babassu coconut fiber retained on the 0.6 mm sieve (P4) with the MOR result for the formulation containing only eucalyptus fiber retained on the same sieve, it can be seen that the results were practically identical. This shows that babassu fiber, because it has a lower basic density, consequently has a slightly higher mass density than eucalyptus fiberboard, produced agglomerated fiberboard that was just as resistant as eucalyptus fiberboard.

The relative differences between the flexural properties of the outer and inner layers are much greater than the difference between the densities of these layers. This makes it possible to conclude that the mechanical properties of the layers are not proportional to their densities. [21]

Figure 8 shows a comparison between the mass density and the flexural modulus of rupture of the formulations studied, using only the values corresponding to the central region of the board. This study also showed that the density had no influence on the flexural modulus of rupture of the panels made with either eucalyptus fiber or babassu coconut fiber.

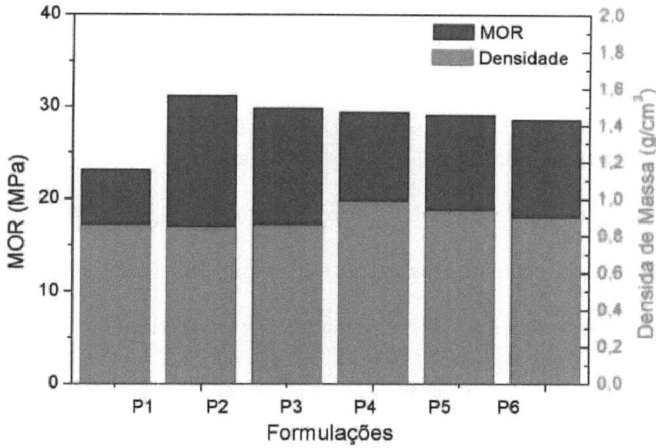

Figure 8 - Comparison between Mass Density and MOR in Flexion

Comparing the average values in Table 11 with studies published by other researchers on particleboard produced with sawmill residues (sawdust and shavings) of Pinus elliottii wood, we find values between 9.74 and 13.58 MPa for MOR. [7]

An important factor to consider in mechanical properties is the compatibility between the type of resin and the characteristics of the fiber.

In static bending tests with UF-shrinked pine fibers, an average of 26.3 MPa is obtained for MOR. For eucalyptus fibers, the average MOR was 25.7 MPa. Using pine fibers shrink-wrapped with inorganic resin, the average MOR was 20.2. For eucalyptus fibers, the average MOR is 19.86 MPa[9].

In general, the panel with eucalyptus fibers (P1) showed the lowest values for MOR, which can be explained by the larger size of the fiber, which may cause less adhesion between the particles. However, considering the characteristics of the particles, more detailed studies could be carried out, looking at factors such as the slenderness ratio, the flatness ratio and the presence of extractives in the residues, in order to better explain the values obtained for this mechanical property.

6   CONCLUSIONS

The thickness swelling values at 2 hours obtained in this study are in the 2% to 6% range. These values are below the minimum value specified by the NBR 14810-2 standard.

Comparing the thickness swelling values of this work with those of the standards and those found by the researchers cited, it is considered that eucalyptus panels, babassu epicarp panels, or panels with 50% eucalyptus and 50% babassu epicarp performed well.

In this work, the values obtained for the 24-hour water absorption of the composites made are in the 15% to 34% range. Thus, comparing the values obtained with the values found in the aforementioned studies, the eucalyptus panels, babassu epicarp panels or panels with 50% eucalyptus and 50% babassu epicarp performed well in terms of water absorption.

The density values obtained in this work are in the 850 kg/m^3 to 990 kg/m^3 range, which is higher than the established standard (551 kg/m^3 to 750 kg/m^3). Therefore, the panels made are classified as high density (HDF).

The values obtained in the Bolt Pullout test range from 1040.06N to 1906.77N. The NBR 14810-2 standard recommends a minimum value of 1020N for this test.

The values obtained for the screw pull-out resistance for each composite were higher than those in the studies and articles cited.

The flexural strength values obtained are in the range of 23 MPa and 31.16 Mpa, well above the values specified by NBR 14810-2 (10MPa to 18MPa) and NBR 15316 (20 Mpa).

The fiber from the epicarp of babassu coconut has technical potential for application in cold-cured wood fiber sheets.

7   REFERENCES

[1] ABNT - ASSOCIAÇÃO BRASILERIA DE NORMAS TÉCNICAS. **NBR 15316-3: Medium-density fiberboard. Part 3: Test methods.** Rio de Janeiro, 2006

[2] ALVITE, J. D. B.; VAZQUEZ, M. C. T.; INFANTE, F. S. **Manual de la madera de eucalipto blanco.** In: Tablero de fibras. Editora Fundación para el Fomento de la Calidad Industrial y Desarrollo Tecnològico de Galicia, San Cibrao das Vinas, Spain, 2002. Chap. IV p.39-50. Available at: <http://www.cismadeira.com/especies/manualeucalipto/capitulosmanual/capitu   os.html>. Accessed on: January 20, 2016.

[3] ASSOCIAÇÃO BRASILEIRA DA INDÚSTRIA DE MADEIRA PROCESSADA MECANICAMENTE - ABIMCI. Sector study 2009, base year 2008. ABIMCI, Curitiba - PR, 2009.

[4] National Bank for Economic and Social Development - BNDES. **Sector: Market overview: wood panels.** Rio de Janeiro, n. 32, p. 49-90, Mar. 2010.

[5] BONDUELLE, A.; YAMAJI, F.; BORGES, C. C. **Residuos.** Wood Magazine. V. 12 Issue 68, December 2002. Available at: <http://www.remade.com.br/br/revistadamadeira_materia.php?num=272&subject= Res%EDduos&title=Res%EDduos>. Accessed on: January 20, 2016.

[6] BRAZIL. Ministry of Agrarian Development, Ministry of Social Development and Fight against Hunger & Ministry of the Environment. **National promotion of the babassu coconut value chain.** Brasilia: Ministry of Agrarian Development, 2009.

[7] BRITO, Edvà Oliveira et al. **Properties of boards produced with coconut fruit residues and pine particles.** Floresta e Ambiente. v. 11, n. 2, p. 1-6, Aug./Dec. 2004.

[8] BUAINAIN. A. **Wood production chain.** Ministry of Agriculture, Livestock and Supply, Secretariat for Agricultural Policy, Inter-American Institute for Cooperation on Agriculture. - Brasilia: IICA: MAPA/ SPA, 2007.84 p.; 17.5 x 24 cm (Agribusiness; v. 6).

[9] DA COSTA, Lourdes Patricia Elias et al. **Physical properties of particleboard made from mechanical wood processing residues of Pinus elliotii Engelm.** Ciência Florestal, Santa Maria, v. 15, n. 4, year 2005, p. 421-429.

[10] ELEOTÉRIO, J. R. **Physical and mechanical properties of MDF panels of different densities and resin contents.** 2000. 120f. Dissertation (Master's Degree in Forestry Engineering) - Escola Superior de Agricultura "Luiz de Queiroz", Piracicaba, 2000.

[11] EUROPEAN COMMITTEE FOR STANDARDIZATION. European Standard EN 622. **Fiberboard - Specifications. Part 5: Requirements for dry-process fiberboard.** Brussels, 2006.

[12] IWAKIRI, S. **The influence of processing variables on the properties of particle boards from different Pinus species.** 130 f. Thesis (Doctorate in Forestry Sciences) - Federal University of Paranà, Curitiba, 1989.

[13] IWAKIRI, S. et al. O. **Production of high-density particleboard using melamine-urea-formaldehyde resin.** In: Revista Cerne. Lavras, v. 11, n. 4, p. 323-328, Oct./Dec. 2005.

[14] MALONEY, T. M. **The family of wood composite materials.** Forest Products Journal, v. 46, n. 2, p. 19-26, Feb. 1996.

[15] MOSLEMI, A. A. **Particleboard.** Illinois: Southern Illinois University Press, 1974. v. 1. 244 p.

[16] NBR 15316-2: **Medium-density fiberboard. Part 2: Requirements.** Rio de Janeiro, 2006.

[17] PEREIRA, Léo C. O.; TAKAHASHI, Rafael; OLIVEIRA, Dênio R. C. de. **Characterization of wood waste and sisal fibres for the manufacture of polyester matrix composite materials.** Congresso Brasileiro de Educaçao em Engenharia,39.Anais.Oct.2011.

[18] SANTIAGO, Francisco Luiz Sanchez. **Study of the technical and economic feasibility of using eucalyptus bark generated in the wood panel manufacturing process.** 102 fls. Dissertation (Master's Degree in Agronomy) - Universidade Estadual Paulista "Julio De Mesquita Filho", Botucatu-SP, 2007.

[19] SANTOS, Washington Luis França. **Using tauari wood waste (couratari oblongifolia) to produce particleboard using castor oil-based resin (Ricinus communis L.).** 97 fl. Dissertation (Master's Degree in Materials Science and Engineering) - Federal Institute of Education, Science and Technology of Maranhao. Sao Luis, 2010.

[20] SILVA, Antonio Joaquim da. **Extractivism of babassu coconut (orbignya phalerata, mart.) in the municipality of Miguel Alves - PI: paths to sustainable local development.** Master's Degree (Development and Environment) - Federal University of Piaui, 2011.

[21] VITAL, Marcos H. F. **Impacto ambiental de Florestas de Eucalipto.** Revista do BNDES, Rio de Janeiro, v. 14, n. 28, p. 235-276, dec. 2007. Available at: <www.bnds.gov.br/SiteBNDS/export/sites/default/.../rev.2808.pdf>. Accessed on: 25 Jan. 2016

[22] WEBER, Cristiane. **Study on the feasibility of using plywood, MDF and MDP waste to produce agglomerated panels.** Curitiba: [s.n.], 2011.

[23] WILCZYNSKI, A.; KOCISZEWSKI M.; **Bending properties of particleboard and MDF layers.** Holzforschung, Vol. 61, pp. 717-722, 2007 · Berlin · New York. DOI 10.1515/HF.2007.116.